As Aesop watched with curious care
Creatures of the earth and air,
The thought grew always in his mind:
This is a tale of humankind.

CRY WOLF
and other
Aesop Fables

Paintings by Barry Castle

Retold by Naomi Lewis

OXFORD UNIVERSITY PRESS

First published in the United States in 1988 by Oxford University Press,
Inc., 200 Madison Avenue, New York, New York 10016

Published in Great Britain in 1988 by Methuen Children's Books Ltd., London

ISBN 0-19-520710-6

Printed in Hong Kong
by South China Print Co

Barry Castle's original works of art
are exhibited exclusively by Portal Gallery Ltd,
London, England

Contents

Foreword

Aesop's fables have been in circulation for between two and three thousand years. Who was Aesop? The most reliable answer comes from the Greek historian Herodotus, writing in the 5th century BC. From him we gather that Aesop the fable-maker, was already a widely known character, that he lived a century earlier (the 6th century BC), that he had been a slave or servant on the island of Samos, in the household of a Samian citizen called Iadmon. Later, it seems, when on some mission, he was killed by the people of Delphi (Plutarch says that he was hurled from a rock) – why, we are not sure. It is possible that the priests and superstitious people feared his wit and intelligence. But his fables had already made him famed through Greece and beyond, and that fame had never grown less. *Was* he ugly, a stammerer, as later chroniclers have guessed? Maybe so; maybe no. Some scholars think that these are fantasies. Facts after all, are few. What is certain is that he had a sharp mind, and that the fable itself, as we know it, was probably his invention. True – not all of the fables known as Aesop's are actually his, for others came to be added over the years. But the basic manner remains.

What is a fable? A short simple moral tale, you could say, in which animals usually carry the burden of the lesson. At best, it makes its point so vividly that the picture and the idea are for ever fused in the listener's or reader's mind. Wolf! Wolf! Dog in a manger! Sour grapes! And there's that tale of the frogs who asked Zeus for a king. He gave them a log, which lay in the water and did not trouble them. A group of dissatisfied frogs demanded a more active ruler. Zeus sent them – some say a stork, some a heron, some a giant eel. Whichever it was, it rapidly ate them up. A lesson here all right, as apt as ever it was. Even briefer, but no less to the point is one in which a wolf observed some shepherds eating a haunch of mutton. 'Ah', said he, 'if I were doing as you do now, what a hullabaloo would be raised against me!'

Still it would be wrong to say that every fable is a good fable. (The same holds for folk tales). Many, reflecting their times and climes, have a streak of mean injustice which thoughtful readers today, (children as much as adults) do well to question. Greed, vanity, laziness, cruelty are human rather than animal qualities; certainly they do not belong to the toiling ass, donkey, camel or mule in the stories or in life. It is interesting to find that Kenneth Grahame, no less, writing on Aesop's fables, gently reproaches the 'moral cowardice' of loading the sins on the 'friendly, tactful, unobtrusive beasts around. The moment they are really studied,' he continues, 'these animals are seen to be so modest, so mutually helpful, so entirely free from vanity, so tolerant and uncomplaining, that a self-respecting humanity would not stand the real truth for a moment. But one *could* deal out the more prominent of human failings among them.' This brings us back to the *reason* of the tales.

All through the historied centuries they have been told and retold, in numerous languages, but their facts have stayed unchanged. The moral comments though are another matter. These were added at different times, by different hands, and very odd, irrelevant or absurd some of them seem now. Usually, indeed, they seem to voice the practical rather than ethical codes of their society. But there is more than one way of judging correct behaviour. The child whose sole response to the parable of the Prodigal Son was, 'Oh, poor fatted calf!' had perhaps reached the best of truths in that always vexing story. Aesop was no fool, and I cannot think that he would have meant his tales to be museum pieces. So, though the texts in this book keep to the classic detail, it seemed right to look again at the 'morals'. Why, for instance, should a man who had risen from servitude put out a fable such as that of the Eagle and the Tortoise (see within) which seems to condemn ambition, imagination, what you will.

I imagine Aesop rising to the challenge and offering a range of interpretations, all of interest to anyone reading now. Indeed, in all the fables here I have looked for the spark of thought that can make the tale as valid for us today as for twenty-five centuries ago. Seen in this light, the original teller, that enigmatic character from the antique past, is not so far away.

Naomi Lewis

Only hope was left

You may ask (said Aesop) what is the best of gifts – wealth? freedom? love?

Wait! There is an ancient story. Zeus collected all the good things in life, packed them into a jar and sealed the lid. Then, knowing the nature of man, he handed the jar to a mortal, bidding him keep it safe until required, but on no account ever to break the seal. Well, of course in time he did; that's human nature. So all the good things escaped but one, and that was Hope. Since then, no human has had *all* the gifts, but the fortunate ones have had hope.

This is also told (continued Aesop) in a different way. A box was given by Zeus to a girl called Pandora, but this contained all the world's ills and evils, and a single good thing, Hope. Pandora waited scarcely an hour before lifting the forbidden lid. Out came all the dreadful contents. They have been around ever since, but Hope is with us still.

MORAL   **Wily Zeus enjoyed his jest**
Knowing his gift worth all the rest.
Hope lights the dark, unlocks the cell,
Makes human living possible,
And prompts the poorest wretch to say,
Tomorrow brings another day.

Men and lions

It chanced that a youth and a lion were travelling along the same road; to while the time each boasted of his skills and powers. Presently they passed a stone carving of a man strangling a lion. The youth, who had been casting about in his mind for ways to impress his great companion, pointed to the statue with a triumphant finger. 'There!' he said. 'You see that men are stronger than lions!'

The lion smiled. 'If,' he observed, 'a lion chose to carve in stone that statue would be showing the man on the ground crying out for mercy to the lion.'

MORAL **Most arrogant of beasts is man.**
'I am the only gifted one!
All else are worthless fools,' says he
To justify his tyranny.
Loudly he proclaims the lie.

What will he do with history?

Thus Aesop in his century
At least five hundred years BC.

Slow but sure

'Poor fellow,' said Hare to Tortoise. 'How can you endure to be so still and slow? What can you see of the world?'

Tortoise said, 'You are quick enough, I grant you, leaping and dancing, skittering here and there. But where does it all lead? As for me, I may be slow, but where I mean to go, I go. Now here's a thought. Give us the same goal and one of us will reach it first. But will that one be you? Maybe yes, maybe no.'

'Oh, oh,' said Hare. 'See that tree at the end of the field? We'll race to it. Start now. You'll find me waiting there.'

Off went Tortoise. The ground was rough, but he kept a straight line, never turning aside. Hare smiled. No hurry. Thoughts of the victory made him prance and dance; then he sat down for a rest.

> He slept. He woke. Unwise to boast.
> The least may well become the most.
> For waiting there – no dream, no ghost –
> Sat Tortoise at the winning post.

MORAL
> Now shift to humans. Here are two
> (said Aesop) first, this lively one,
> with every hour a notion new,
> a plan, a scheme. True, nothing's done,
> nothing is followed through.
> The other keeps one goal in view,
> avoids diversion, glumly plods
> to invent the wheel or crack the clue,
> And does it too against all odds.
> (You'd think he'd aggravate the gods)
>
> Now who is who?
> Which one am I?
> Which one are you?

A waste of good counsel

One came to Aesop saying, 'Well, master-servant*, here is a tale. Now what do you make of it?'

This was the tale:

A tortoise with a restless mind longed for a different view of the world. 'I know it only at earth level, a few steps here, a few steps there, and *that* for me is a day's journey.' He begged the eagle to teach him how to fly.

'A foolish notion,' said the bird. 'Quite preposterous. You are not made for flight.'

'Then carry me in your claws, so that I can see the world from above, even if only this once.'

The eagle picked up the tortoise then, at a great height, let him drop, to perish on the rocks below.

Aesop's reply I could read this several ways,
Some of blame and some of praise.
No single answer fits the case.
The lowly born should know their place?
You cannot wish beyond your range?
It's all ordained; things cannot change?
Or – greatly wishing is no vice?
But every wish demands its price?
No vision can be wholly lost?
The dream fulfilled is worth the cost?

*Aesop was reputed to be a slave.

City of lies

In a desert waste a traveller came on a woman sitting alone, her back to the distant town. He said, 'Who are you? Why are you waiting in this desolate place?'

'I am Truth.' she said. 'This is my home.'

'Then why are you not where Truth should be, with humans in the city?'

Because, wherever humans are many, liars are many also. When humans still were few, then lies were few. But in cities of today I cannot live.

MORAL   Humankind's unchecked increase,
Leave no life for Truth or Peace
Creeds are now to monsters grown.
Each claims Truth to be its own.
Who seeks for Truth must seek alone.

The victor vanquished

Two cocks fought for the favors of certain hens. The loser
went and hid in a dark corner. But the winner perched on a
high wall and boasted of his victory at the top of his voice.
Cock-cock-cock-a-doodle-doo! The noise echoed everywhere.

An eagle heard the foolish cock, swooped down and carried
him off – a tasty dish! The loser waited a while, came out at
last, and the winner's place was his.

MORAL   **The boaster loses what he gains**
His noise annoys; his favor wanes.
Bide your time with secret face
You yet may take the boaster's place.

One swallow

A wild young man, used to having whatever he wished, was left a large fortune when his father died. He spent it fast, on gambling, feasts and other pleasures, and soon he had nothing left but the clothes upon his back. One fine morning in early spring he saw a swallow, a young bird which had lost its way and had landed alone, before the proper time.

'Ah,' said the young man, 'A swallow! That means good weather, so they say; summer will soon be here.' And he went off and sold his cloak. But spring is treacherous; days came of frost and icy winds, and the young man, shivering, found the untimely swallow frozen to death.

'Wretched creature!' he cried. 'You have destroyed yourself and me.'

MORAL   **Several morals I observe.**
The feckless get what they deserve.
Hasty actions, lacking thought,
Bring disaster in their court.
Superstition is a snare.
Don't go by hearsay; use your mind.

Swallows too should have a care.
Swallows, stray not from your kind.

A voice and little more

A lion, half asleep was roused by a croaking sound. Quarck! Quarck! 'What is that mighty animal?' he mused. 'On my territory, too.'

He strolled off to investigate, and found – a frog. 'Midge of a creature!' he said. 'Paltry atom! What have *you* to say that warrants so large a noise?' He flicked it up and crushed it with his paw.

MORAL   **In human terms this would suggest**
A prudent silence can be best.
What room is there for worth within
When smallest minds make loudest din?
No space remains for thought or doubt.
The great mind has no need to shout.

This is, of course (said Aesop then)
Unfair to frogs, but true of men.

The fault, dear Brutus, is not in our stars, but in ourselves

A man who had journeyed far on foot saw a well and stopped to drink. Then, tired out, he sat for a while on the edge, and there fell asleep. He would have tumbled in if Fortune had not passed and woken him. She shook a reproachful finger. 'You speak about me often,' she remarked. 'You call me Luck or Chance, and blame me for your troubles. Now, if you had fallen in the well, would you have said the fault was yours or cried, 'Misfortune!'? I think I know the answer; so do you.'

MORAL   **Luck watches humans toil and play,**
                    Likes the alert one, goes half-way.
                    They meet. All's well.
                    But (as the legends also tell)
                    The one who neither thinks nor tries
                    Misfortune snatches as her prize.

Cry wolf

A shepherd boy saw a movement in the bushes. 'A wolf!' he thought. He rushed to a little hill and called for help – 'A wolf! a wolf!' At once a crowd of villagers left their work and ran to save the flock. This made the boy laugh, and a few days later he again called 'Wolf!' just to see the villagers come running. He played the trick a third time, and a fourth. But the people had grown tired of being made to look like fools and paid no more attention to his shouts. So when a wolf really did appear the boy cried 'Wolf!' in vain.

Nobody came to help; sheep were lost, and so was the shepherd's job.

MORAL   **Don't play with truth, I tell you true,**
or truth in turn will play with you.

27

A friend in need

Two friends were traveling through a wood when a bear lumbered towards them. One man hastily climbed a tree and hid among the leaves. The other, short of time, lay on the ground and held his breath, pretending to be dead. 'I've somewhere heard,' he told himself, 'that a bear won't touch a corpse.'

The creature had no ill-intent. It sniffed him over, thought its thoughts, then wandered on its way. Now the first man climbed down. 'That bear was whispering in your ear,' he said. 'Tell me, what did he say?'

'It is interesting that you ask,' said the other. 'He advised me not to travel with friends who desert in time of danger.'

MORAL   **Need verse explain? The moral's plain.**
The fine day sends
In droves, new friends –
But which stay in the rain?
No riddle's here. The lesson's clear –
Which friends are gold, which disappear –
Just read the tale again.

Plucked clean

A man in middle life, whose hair was turning silver, was watched with sullen eyes by his wife.

'Husband!' said she, 'the dark hairs in your head make you appear too young. I'll take them out.'

'As you will,' said he. 'All I ask for is a quiet life.'

His daughter presently looked at him, her head on one side, then on another. 'Father,' she said, 'your white hairs make you look too old. I'll take them out.'

Before long he was bald, pleasing to neither nor yet to himself.

MORAL   Please all, please none.
         My lesson's done.
         Who reads may run.